PTSD And CBD Oil

Understanding The Benefits Of Cannabis And Medical Marijuana

Jane Fields

Table of Contents

Legal Notice

In no way should the information in this book be considered medical advice. It is provided for information purposes only. Never disregard medical advice or delay in seeking it because of something you have read here or on a website.

The author assumes no responsibility for the improper use or self-diagnosis and/or treatments using the information in this book. The information should not be confused for prescription medicine. The information should not be used as a substitute for medically supervised therapy. If you suffer from clinical nutrient deficiencies, consult a licensed, qualified doctor.

Before starting a dietary supplement, always check with your medical doctor to prevent interactions with any medications you are on or to prevent overdosing. This is particularly important for people who are breast-feeding, chronically ill, under 18, or taking prescription or over the counter medicines. Certain supplements can boost blood levels of certain drugs and/or micronutrients to dangerous levels. None of the information below is intended to be a treatment protocol for any disease state, but rather are offered to provide information and choices regarding nutritional support. None of the information is intended to be construed as medical advice in this book.

No action should be taken on your condition solely based on this books content, the perceived scientific merit of the author or this book. Readers who fail to consult their physician prior to using any information from this book assume the risk of any adverse effects. The use of nutritional supplements for any reason, other than to increase dietary intake levels of specific nutrients is neither implied nor advocated by the author.

The results reported may not occur in all individuals. This is a comprehensive limitation of liability for damages arising out of or in connection with the use of this book of any kind, including (without limitation) compensatory, direct, indirect or consequential damages, loss of income, profit, loss or damage to personas for property claims by third parties, The below is not intended to diagnose, treat, cure or prevent any disease.

The below information is provided for educational and information purpose only and is not intended to be a substitute for a health care provider's consultation. The information in this book should not be relied upon to suggest a course of treatment for any particular individual. It should not be used in place of a consultation with your physician or other qualified healthcare professional.

INTRODUCTION

The plant Cannabis sativa has been used in medicinal practice for thousands of years. Cannabis is a complex plant with over 400 chemical entities of which more than 60 of them are cannabinoid compounds, some of them with opposing effects. The pharmacologically active constituents of the plant are termed cannabinoids, phytocannabinoids (cannabinoids derived from the plant), synthetic cannabinoids (artificial compounds with cannabinomimetic effects), and endocannabinoids (endogenous compounds with cannabinomimetic effects) all of which act together on the human endocannabinoid system (ECS), which regulates various functions in the body.

The roots of Cannabidiol (CBD), specifically the seeds and cannabis oil were used for food in China as early at 6,000 BC. Today, individuals with simple to chronic disease, organic and health conscientious supplement users and alternative medicine seekers are more interested in the health-related properties of the Cannabis compounds in the fight against chronic conditions and debilitating disorders including Acne, Arthritis, Cancer, Crohn's Disease, Epilepsy, Fibromyalgia, Glaucoma, Hepatitis C, HIV/AIDS, Multiple Sclerosis, Nausea, Parkinson's Disease, PTSD - Post Traumatic Stress Disorder, Schizophrenia and everyday Stress are turn to a new found (but ancient) natural medicine in their fight to cure themselves. Those with life-threatening diseases, to those in chronic pain and those with common aliments are seeking the benefits of CBD based Oils and products derived from Cannabis to help alleviate conditions, aid in healing and increase general well being.

In the United States, Cannabis Sativa is a Schedule I substance and its use for recreational or medicinal means is illegal according to Federal law. However, contrary to Federal policy, individual state laws have allowed for medical use of marijuana in 29 states[1] and recreational use in 9 states; including Washington D.C.

Given the evolving policies regarding the medical use of cannabis, physicians are increasingly prompted with questions about its therapeutic role for a variety of disorders. Research, studies and science are now able to prove the medicinal benefits of CBD and THC oil, and finally offer natural alternatives to pharmaceuticals and traditional prescriptions. Consumers and patients both can now fight conditions and aliments by turning to CBD Oil which is natural, plant based, opioid free, healthier and better aligned to an organic chemical free lifestyle.

---

[1]http://medicalmarijuana.procon.org/view.resource.php?resource ID=000881

CHAPTER ONE: WHAT IS CBD OIL?

Cannabis oil, known as CBD oil, is derived from the seed, stalk and flowers of the cannabis sativa plant and has a significant amount of the compound Cannabidiol. Cannabidiol is one of the more than 111 *active* cannabinoids identified in the cannabis sativa plant. CBD oil is the result of utilizing one of numerous extraction methods to isolate, preserve and maintain the purity of the medicinal resin (oil or residue) that is found on the flowering leaves of the plant.

CBD and CBD oil independently is not psychoactive. This means that it does not change the state of mind of the person who uses it. However, it does appear to produce significant changes in the endocannabinoid system (ECS) system of the human body, and it is proving to have a long list of medical benefits. Most of the CBD used in today's consumer products is found in the least processed form of the cannabis plant, known as hemp. Hemp is different than CBD although derived from the same plant.

For over 40 years researchers have been looking at the potential therapeutic uses of cannabinoids and the different concentration levels along with the interactive properties of CBD and THC, independent and collectively. Until recently, the most well-known compound in the cannabis plant was THC, or Delta-9 Tetrahydrocannabinol. THC is the most active ingredient in Marijuana, which is also found in the Cannabis Sativa plant. Marijuana contains both THC and CBD, but the compounds have different effects. THC is well-known for the mind and body "high" it produces when ingested, such as when smoking the plant or when using oil in edible form, such as for cooking into foods.

CHAPTER TWO: WHAT ARE CANNABINOIDS?

Cannabinoids are naturally occurring compounds found in the Cannabis sativa plant. Over 480 different compounds are present in the plant, but only around 60 are termed *active* cannabinoids. The most well-known among these compounds is Delta-9-Tetrahydrocannabinol (9-THC), which is the main psychoactive ingredient in cannabis. Cannabidiol (CBD) is another predominant and important compound present, which makes up about 40% of the plant resin extract.

For the purpose of this book, we will focus on specific cannabinoids, which are separated into the following subclasses:

Cannabidiol (CBD)

Cannabichromene (CBC)

Cannabigerol (CBG)

Cannabinodiol (CBDL)

Cannabinol (CBN)

Tetrahydrocannabinol (THC)

Differences between Cannabinoids

The main way in which the cannabinoids are differentiated is based on their degree of psycho-activity.
For example, CBG, CBC and CBD are not known to be psychologically active agents whereas THC, CBN and CBDL along with some other cannabinoids are known to have varying degrees of psycho-activity. The most abundant of the cannabinoids is CBD, which is thought to have a wide range of benefits, one being anti-anxiety effects, which are possibly counteracting the psychoactive effects of THC when in the same strain.

When THC is exposed to the air, it becomes oxidized and forms CBN which also interacts with THC to lessen its impact. This is why cannabis that has been left outside of an air-tight container, will has less potency and residual effects when smoked due to the increased CBN to THC ratio.

Effects of Cannabinoids

Cannabinoids exert their effects by interacting with specific cannabinoid receptors present on the surface of cells throughout our body. These receptors are found in different parts of the human central nervous system and the two main types of cannabinoid receptors are referenced as CB1 and CB2.

Laboratory research[2] in 1992, yielded findings of a naturally occurring substance in the brain that binds to CB1, Anandamide. Although discovered decades previous the relation to the CB1 receptors was not understood. This cannabinoid-like chemical and others that were later discovered are referred to as endocannabinoids.

The effects of cannabinoids depends on the area of the brain which are involved. Effects on the limbic system may alter the memory, cognition and psychomotor performance; effects on the mesolimbic pathway may affect the reward and pleasure responses and pain perception may also be altered.

---

[2]https://www.researchgate.net/publication/6126760_Discovery_and_Isolation_of_Anandamide_and_Other_Endocannabinoids

Δ-9-tetrahydrocannabinol (THC)

Primary Cannabinoids in Cannabis

THC

THC is the abbreviation for both Delta(9)-Tetrahydrocannabinol and Delta(8)-Tetrahydrocannabinol, which is the lesser psychoactive THC cannabinoid. As most cannabis lovers probably know, THC is the primary psychoactive compound in marijuana. The list of medical conditions of which THC is found to have benefit is rapidly growing, so for purpose of this book we will focus on a few of the notable conditions with published studies and research:

- Alzheimer's Disease
- Neuropathic pain

- Multiple Sclerosis
- Parkinson's Disease
- PTSD
- Cancer
- Crohn's Disease
- Chronic (pain) relief

Some resources claim that Delta(8) may have neuroprotective and anti-anxiety properties, making it an interesting companion to the more notorious psychoactive. However, more research is needed to confirm just how this particular compound acts inside the body. Here are a couple of additional benefits you can expect from Delta(8)-THC:

- Anti-Anxiety
- Appetite Stimulator
- Pain Relief
- Neuroprotection
- Anti-Nausea

CBD

CBD, which is short for Cannabidiol, is the second most famous cannabinoid. Like THC, the list of medical benefits of this cannabinoid just keeps getting longer. Unlike THC, CBD is non-psychoactive. It's also now legal in more states than its more controversial counterpart and available in a large variety of products and forms. Once mainstream media learned that CBD had medical value, a whirlwind of cannabis research ensued. Here are a just a handful of conditions CBD can treat:

- Acne
- Anxiety
- Arthritis
- Cancer
- Depression
- Diabetes
- Epilepsy
- Fibromyalgia
- Glaucoma
- Hepatitis C
- Psychotic Disorders
- Chronic Pain
- Multiple Sclerosis
- Nausea
- Parkinson's Disease
- PTSD Post Traumatic Stress Disorder
- Schizophrenia
- Stress

CBC

Cannabichromene (CBC) is an abundant naturally-occurring phytocannabinoid, and is thought to be the second most abundant cannabinoid in cannabis. CBC has been shown to produce anti-nociceptive (painkilling) and anti-inflammatory effects in studies. CBC shares the same molecular formula as THC and CBD: $C_{21}H_{30}O_2$. Although many cannabinoids share the same formula, the atoms within the molecule are arranged in slightly different ways. In some strains, CBC may even take dominance over CBD. Similar to

CBD, Cannabichromene is non-psychoactive.

Here are benefits of CBC rich strains:

- Encourages Brain Growth
- Anti-Inflammatory
- Anti-Depressant
- Anti-Bacterial and Anti-Fungual
- Pain Relief

CBN

Cannabinol, emerges when the dried flower has gone a bit stale. THCa breaks down into this compound over time and results in a mildly different compound. CBN's most pronounced, characterizing attribute is its sedative effect, and studies are showing, 5mg of CBN is as effective as 10mg dose of diazepam, a mild pharmaceutical sedative. If you leave some bud out sitting out in the open air for too long, you'll eventually have a product with larger amounts of CBN. CBN's studied benefits include:

- Appetite Stimulant
- Anti-Inflammatory
- Anti-Bacterial
- Pain reliever
- Anti-Asthmatic
- Anti-Insomnia
- Potential Medication for Glaucoma

CBG

Cannabigerol (CBG) is the building block for many compounds in marijuana, including THC and CBD. This cannabinoid is found early on in the growth cycle, which makes it somewhat difficult to find in large quantities. In fact, THC, CBD and many other cannabinoids all begin as CBG. CBG is non-psychoactive and can be thought of as the "stem cell" or "parent" of other cannabinoids. After being synthesized, CBG is quickly converted to other cannabinoids through natural processes that occur within the cannabis plant. This explains the low CBG content of most cannabis strains. Research and studies are finding CBG benefit in supporting:

- Anxiety and Pain
- Glaucoma
- Anti-Inflammation
- Skin Conditions
- Digestive Conditions
- Anti-Septic
- Neuroprotection
- Cancer
- Huntington's Disease

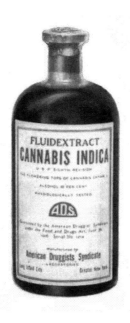

CHAPTER THREE: HISTORY AND
MISCONCEPTIONS OF CBD OIL

Cannabis (Medical marijuana) has long been used to treat a variety of ailments and conditions since ancient times. However, CBD oils and related products are now a prevalent topic in today's medical field with continued discoveries of the profound effect on children and adults with debilitating and chronic disease.

Using CBD dates back to the 19th century and is noted Queen Victoria would use cannabis to alleviate menstrual cramp pain. CBD oils and related products until recently have always played second to its fellow cannabinoid THC. In the 1980s, studies hinted that

CBD could alleviate certain types of pain, anxiety and nausea, but still did not get much attention until the 90s.

In 1998, a medical company called GW Pharmaceuticals based in England began to cultivate cannabis specifically for medical trials. Their aim was to develop a concise and consistent plan for extracting CBD. Geoffery Guy[3], one of the company's founders, staunchly believed that cannabis plants rich in CBD would be used as medicine. This research would lead to scientific studies conducted by the International Cannabinoid Research Society[4], Society of Cannabis Clinicians[5], and International Association for Cannabinoid Medicine[6]. Early studies showed that CBD lessened anxiety and reduced the frequency and severity of seizures. This turned heads in the medical community, and cannabis strains began to be cultivated with extremely low levels of THC and high levels of CBD.

In the 2010s, the public began to see what a profound effect CBD oil could have treating a variety of life-threatening ailments, especially in children. One of the most encouraging and perception changing stories is that of Charlotte's Web[7], which is named after

[3] https://www.gwpharm.com/about-us/board-directors
[4] http://icrs.co/#About
[5] http://cannabisclinicians.org/
[6]https://www.cannabis-med.org/index.php?tpl=page&id=73&lng=en

[7] https://en.wikipedia.org/wiki/Charlotte%27s_web_(cannabis)

Charlotte Figi[8], born October 18, 2006. Her story has led to her being described as "the girl who is changing medical marijuana laws across America." Her parents and physicians say she experienced a reduction of her epileptic seizures brought on by Dravet syndrome after her first dose of medical marijuana at five years of age. Her usage of Charlotte's Web CBD Oil was first featured in the 2013 CNN documentary "Weed". Media coverage increased demand for Charlotte's Web and similar products high in CBD, which has been used to treat epilepsy in toddlers and children.

In the modern era, advances in chemistry have allowed people to consider both raw CBD oil, closer to its minimally filtered historical form, and more refined extractions that increase the levels of CBD. Moreover, developments in technology and extraction have created a variety of formulas and products like dermal skin patches, lotions, balms, and a wide variety of edibles. Today, the options in CBD oil supplements are so vast the first farmers could have never dreamed of, but of course, this also requires that each person carefully investigate the supplements that are right for them condition and desired benefit.

---

[8] https://en.wikipedia.org/wiki/Charlotte%27s_web_(cannabis)

Misconceptions

Cannabis has gotten a wave of new momentum following Dr. Sanjay Gupta's documentary regarding the medicinal properties and benefits of cannabis. Cannabidiol - or CBD - has been given the title of being medicinal because it treats without giving a person the signature high cannabis has been known for. THC or Tetrahydrocannabinol is that cannabinoid or the non-medical type which has gotten much more attention in the past. Both of these play important roles in medicinal marijuana treatments, but there are several misconceptions ,which need to be cleared up.

From medical marijuana centers, to cutting edge medical teams, and even across major news outlets, it's a buzzword that is only growing in popularity and interest. While CBD does not get you "high," like its cousin THC, it has certainly piqued the interest of everyone from seasoned marijuana growers to industrial hemp farmers to esteemed medical teams. With great buzz, comes information and a myriad of "facts" claiming that it is a cancer-curing, seizure-stopping miracle oil. While there are certainly a large number of cases cited with patients who CBD and CBD Oils have helped change or improve their conditions significantly, it is imperative to separate what we know as scientific evidence from fallacies about the cannabinoid. Plenty of start-ups and e-commerce retailers have jumped on the CBD bandwagon, touting CBD derived from industrial hemp as the next big thing, a miracle oil that can shrink tumors, reduce or stop seizures, and ease chronic pain without making people feel "stoned." But along with a growing awareness of Cannabidiol as a potential health aid there has been a proliferation of misconceptions about CBD. Here are six of them.

"CBD is medical. THC is recreational."

Often people say they are seeking "CBD, the medical part" of the plant, "not THC, the recreational part" that gets you high. Actually, THC, "The High Causer" has a growing list of therapeutic benefits and properties. Scientists at the Scripps Research Center in San Diego reported that THC inhibits an enzyme implicated in the formation of beta-alkaloid plague, the hallmark of Alzheimer's-related dementia. The federal government recognizes single-molecule THC (Marinol) a synthetic man-made THC structurally similar to its natural counterpart but dissolved in sesame oil, as an anti-nausea compound and appetite booster, deeming it a Schedule III drug, a category reserved for medicinal substances with little abuse potential. Ironically, the whole marijuana plant, the only natural source of THC, continues to be classified as a dangerous Schedule I drug with no medical value.

"THC is the bad cannabinoid. CBD is the good cannabinoid."

The Cannabis Hater strategic retreat: Give ground on CBD while demonizing THC. Diehard marijuana haters are exploiting the good news about CBD to further stigmatize THC in cannabis, casting Tetrahydrocannabinol as EVIL, whereas CBD is framed as the GOOD. Why? Because CBD doesn't make you high like THC does.

"CBD is most effective without THC."

THC and CBD are the power couple of the cannabis compound as they work well together. Scientific studies have established that CBD and THC interact in synergy to enhance each others therapeutic effects. British researchers have shown that CBD potentiates THC's anti-inflammatory properties in an animal model of colitis. We encourage each person to discover what works for them and of course if you live in a state where Marijuana is illegal for recreational use then you might want to consider moving to a state like Colorado who has legalized adult use.
"Single-molecule pharmaceuticals are superior to

'crude' whole plant medicinal."

According to the US Government, specific components of THC and CBD have medical value, but the plant itself does not have medical value. Uncle Sam's single-molecule blinders reflect a clear bias that privileges large pharmaceutical products. Single-molecule medicine is the predominant corporate way, the FDA-approved way, but it's not the only way, and it's not necessarily the optimal way to benefit from cannabis therapeutics.

"It matters where CBD comes from."

People think it often takes large amounts of industrial hemp to extract a small amount of CBD but that's simply not the case anymore as many of these industrial hemp plants have been bred specifically to have a high CBD content. The extraction method can also affect how much CBD is extracted per plant as well as its potency. Almost all hemp derived CBD oil is made using CO_2 extraction as opposed to most marijuana extractions that used a harsh chemical solvent. While medical marijuana legislation is opening up and more and more states are planting crops for medical and commercial purposes, it is by no means the only means to extract CBD. In fact, these hemp crops tend to produce almost identical THC:CBD ratios as medical marijuana.

"CBD is legal throughout the United States"

CBD products are illegal in the United States based on Federal Law, if the product contains any amount of THC. A product that contains THC, even in small amounts, is considered Marijuana under federal law and is illegal. When it comes to CBD products such as tinctures, lotions, balms, and beyond, the cannabinoid can be extracted from leaves, flowers, and the stalk, similarly to industrial hemp. While legislation is in motion to ease up on industrial hemp cultivation, there are only a few states that allow the cultivation for commercial, research or pilot programs.

CHAPTER FOUR: HOW CBD OIL IS MADE

Cannabidiol (CBD) oil is derived from the Cannabis stavia plant and extracted specifically from the high-resin glands or *trichomes*[9] found mainly on the plant's odiferous female flowers (the buds) and to a lesser extent on the leaves. There are also the smaller trichomes, which dot the stalk of the hemp plant, but these contain hardly any resin. Non-glandular hairs shaped like tiny inverted commas also cover the plant's surface.

One of the original creation methods and the man who re-invented CBD oil is Canadian, Rick Simpson. Rick Simpson oil (RSO) is highly-concentrated cannabis oil extracted from female cannabis plants that contain at least 20 percent THC or more. Simpson favors high-THC and low-CBD content when treating maladies. RSO was created when he recalled a study published in the Journal of the National Cancer Institute[10] stating that THC was found to kill cancer in mice. Simpson proceeded to apply the oil topically to recently identified cancerous spots on his neck and face and covered them with bandages. In just four days, the skin underneath was healthy and pink. He was healing. Excited that he had discovered a cure to his cancer, Simpson began helping others heal, spreading the good word ever since.

---

[9] https://en.wikipedia.org/wiki/Trichome
[10] https://www.cancer.gov/about-cancer/treatment/cam/patient/cannabis-pdq

Despite encountering opposition from his own doctor, local authorities, pharmaceutical companies, government health agencies, and the United Nations, Simpson has not only healed his own ailments, he's also successfully treated over 5,000 patients free of charge for all kinds of conditions, including cancer, HIV/AIDS, insomnia, diabetes, depression, osteoporosis, asthma, and more. He even cured his mother's psoriasis.

The active CBD compound in the flower can be extracted, purified and concentrated using a limited number of precise extraction and production techniques. When oil of a higher potency and concentration is derived from the glands it can be used in lower dosage. The higher purity and lower dosage can save money as CBD oil is an expensive product to produce at the highest quality and desired compound blend.

Industrial production of CBD oil is done by combining the cannabinoid rich plant with other compounds like CO_2, butane, ethanol or olive oil, of which can leave their residues in the final product. The most important part of CBD oil extraction is selection of right plant for the type of oil extraction. Typically cost of the final product depends on its potency and purity, origin, heritage, along with the grow process that was used. Oil is in liquid form it can be consumed orally without smoking and can be easily administered to children and adults for whom it is prescribed as medication.

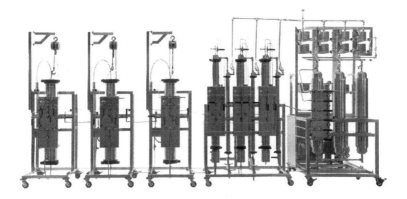

Why is Cannabis extraction performed?

Cannabis extraction can be performed for a variety of reasons, as well as in variety of ways. The most widespread reason, of course, is to obtain potent extracts high potency resin that can be used in the medical marijuana industry to treat patients dealing with a variety of illnesses and symptoms. On a more academic level, cannabis extractions are also performed purely for research purposes in an effort to understand the chemistry of the plant and make scientific inquiries into certain claims.

Just as other compounds might be available in various solid or liquid forms, the effects of cannabis can be achieved in a variety of extract forms. The following is a list of just a few of the most common cannabis extraction methods:

SuperCritical Co2

CO2 extraction machines essentially freeze and compress CO2 gas into a "supercritical" cold liquid state. Carbon dioxide is considered a "supercritical fluid" that becomes a liquid under pressure. In extractions, CO2 leaves no residue, which makes it popular for use in extraction processes across a variety of industries. In this kind of cannabis extraction, supercritical CO2 is put into a pressurized container with the cannabis material. The mixture is then put through a filter to separate the CO2 from the cannabis and, when the pressure in the system is released, the CO2 evaporates back into a gas.

Referred to as "subcritical" or "supercritical" CO2 method, it uses carbon dioxide under different pressures to extract the medicinal oil. By pushing Co2 through the plant at high pressures and low temperatures, CBD can be extracted in its purest form. The CO2 method is considered the most scientific method and this is also among the cleanest techniques of extraction.

This process is often thought of as the best and safest as it cleanly extracts CBD, removing substance like chlorophyll and leaving no residue. CBD oil extracted in this way has a cleaner taste and this is among the most expensive techniques of extraction due to use of hi-tech equipment that must be operated by trained professionals. The advantage of this technique is that the end product is the purest form of CBD oil, which is highly potent and free of chlorophyll and other potential harmful toxins. However, if the heat used during extraction is too high it can damage Terpenes in the oil that have therapeutic benefits and provide flavor and essence to the oil strain.

Distillation

Distillation is probably the most common technique; short path distillation exposes the cannabis oil to heat and vacuum in order to separate a variety of compounds from the extract (cannabinoids and Terpenes, in particular). Using an appropriate laboratory setup, heat is first introduced to the extract to evaporate the cannabinoids and Terpenes. Then, the vapor enters a condenser tube, through which the cannabinoids travel and condense into the recipient flask. This final product does not need to be winterized, as the waxes and fats cannot vaporize and, therefore, remain in the first flask of unprocessed concentrate. This method generally produces the highest concentrations of cannabinoid molecules.

Carrier Oil

The carrier oil extraction method is regarded as most inexpensive method of extracting CBD oil and is recommended by Dr. Arno Hazekamp[11], director of phytochemical research at Bedrocan BV that supplies

---

[11] https://www.researchgate.net/profile/Arno_Hazekamp

it to Dutch Health Ministry. CBD oil extracted by this method will contain a healthy dose of Omega rich acids and minimalistic chemical residues in the pure oil. Usually hemp seed oil or olive oil is used as a carrier in these methods as this is the most effective way to extract resin from the plants and flowers. The only drawback is the short shelf life of oil extracted in this manner though it is highly effective when taken orally or applied topically on the skin.

Live Resin Extraction

Most commonly, cannabis is dead and dried before it is processed. In Live Resin Extraction, the live plant is frozen (such as with liquid nitrogen) right after it is harvested in order to perform extraction. Using a closed-loop hydrocarbon extraction machine this process creates a resin product. The live resin extraction method generally gives a flavor that is more true to the original strain and uses the living plant. This is also one of the most expensive methods to perform because most of the plant's terpenes (essential oils) in the product resulting in a high potency oil.

Terpene Isolation

Terpenes are hydrocarbon oils that are responsible for fragrance in a variety of plants. A wide variety of plant essential oils including those with pine or citrus scents are comprised primarily of terpenes. In the case of cannabis, some experts believe that the terpenes of the plant are equally as responsible as THC for the high experienced during use.

Solvents

Extraction with solvents is a relatively inexpensive method and typically preferred by small-scale producers of CBD and THC oils. During extraction this method uses solvents like butane, ethanol and alcohol derived from grains. This method has several disadvantages the worst of which is potential of explosion while the second is leftover residue of these solvents. Scientists and doctors advice against the use of this method as it can make the end product unsafe for medical use and also can make existing medical condition much worse. When there are unsafe residues in CBD oil it reduces healing powers and can even compromise health of patients.

CHAPTER FIVE: HOW CBD WORKS

All cannabinoids, including CBD, attach themselves to certain receptors in the body to produce their effects. The human body produces certain cannabinoids on its own. It has two receptors for cannabinoids, called CB1 receptors and CB2 receptors.

The CB1 receptor was discovered in 1990, while CB2 was uncovered shortly thereafter in 1993 by a research group at Cambridge University. One source claims that these two receptor types employ significantly different signaling mechanisms. It is known that they are expressed in vastly different ways, including their appearance in various parts of the body (different regions of the endocannabinoid system).

CB1 receptors are present in very high levels in several

brain regions and in lower amounts throughout the body and nervous system. The CB1 receptors, found predominantly in the *cerebellum* and *neocortex* regions of the brain deal with motor coordination and initiation of movement, pain, emotions and mood, thinking, appetite, and memories, among others. THC attaches to these receptors and help mediate many of the psychoactive effects of cannabinoids.

THC has been shown to possess a high binding affinity with CB1 receptors in the brain, central nervous system, connective tissues, gonads, glands, and related organs. This is one reason that consumption of cannabis oils from strains containing a high amount of THC result in a relatively potent effect, giving patients significant relief from pain, nausea, or depression while delivering a strong euphoria to lifestyle users. Those undergoing chemotherapy and patients suffering conditions involving inflammation, like arthritis and lupus, gain significant efficacy.

Think of it like an electrical plug connecting to a wall socket. A THC molecule is perfectly shaped to connect with CB1 receptors. When that connection happens, THC activates, or stimulates, those CB1 receptors. Researchers call THC a CB1 receptor agonist, which means THC works to activate those CB1 receptors. THC partially mimics a naturally produced neurotransmitter known as anandamide, aka "the bliss molecule." Anandamide is an endocannabinoid, which activates CB1 receptors.

When we're talking about cannabis and psycho-activity, we're dealing exclusively with CB1 receptors, which are concentrated in the brain and the central nervous system. The difference between CBD vs THC comes down to a basic difference in how each one interacts with the cannabinoid (CB1) receptor. THC binds well with CB1 cannabinoid receptors. That's where the two diverge.

CB2 receptors, on the contrary, are located throughout the immune system and related organs, like the tissues of the spleen, tonsils, and thymus gland. They are also common in the brain, although they do not appear as densely as CB1 sites and are found on different types of cells.

CB2 sites are also found in greater concentrations (density) throughout the gastrointestinal system, where they modulate intestinal inflammatory response. This is why sufferers of Crohn's disease and IBS gain such great relief from cannabis medicine. It is also a powerful example of how the endocannabinoid system, when supplemented by external cannabinoids (such as from cannabis), can provide such powerful and long-lasting relief for patients of diseases like Crohn's. Cannabis and CBD oil has been shown to have such great efficacy for this condition that, in nearly half of cases, the medicine puts the disease into full remission. The author can personally affirm to the medicinal benefits of CBD and THC oil used in the treatment, remission and chronic pain relief of Crohn's disease as diagnosed in 1983.

Serotonin Receptor

At high concentrations, CBD directly activates the 5-HT1A (hydroxytryptamine) serotonin receptor, thereby conferring an anti-depressant effect. This G-coupled protein receptor is implicated in a range of biological and neurological processes, including (but not limited to) anxiety, addiction, appetite, sleep, pain perception, nausea and vomiting. 5-HT1A is a member of the family of 5-HT receptors, which are activated by the neurotransmitter serotonin. Found in both the central and peripheral nervous systems, 5-HT receptors trigger various intracellular cascades of chemical messages to produce either an excitatory or inhibitory response, depending on the chemical context of the message. CBDA, Cannabdiolic acid, the raw, unheated version of CBD that is present in the cannabis plant, also has a strong affinity for the 5-HT1A receptor (even more so than CBD).

Both CBD and CBDA trigger an inhibitory response that slows down 5-HT1A signaling. In comparison, LSD, mescaline, magic mushrooms, and several other hallucinogenic drugs activate the 5-HT2A receptor, which produces an excitatory response.

Vanilloid Receptors

TRPV is the technical abbreviation for "Transient Receptor Potential Cation" Channel Subfamily V."

TRPV1 is one of several dozen TRP (pronounced "trip") receptor variants or subfamilies that mediate the effects of a wide range of medicinal herbs.

CBD directly interacts with various ion channels to confer a therapeutic effect. CBD, for example, binds to TRPV1 receptors, which also function as ion channels.

TRPV1 is known to mediate pain perception, inflammation and body temperature. Scientists also refer to TRPV1 as a "Vanilloid receptor," named after the flavorful vanilla bean. Vanilla contains Eugenol, an essential oil that has antiseptic and analgesic properties; it also helps to unclog blood vessels.

CBD is a TRPV1 "agonist" or stimulant. This is likely one of the reasons why CBD-rich cannabis is an effective treatment for neuropathic pain.

GPR55 Orphan Receptor

Whereas Cannabidiol directly activates the 5-HT1A serotonin receptor and several TRPV ion channels, some studies indicate that CBD functions as an antagonist that blocks, or deactivates, another G protein-coupled receptor known as GPR55.

GPR55 has been dubbed an "orphan receptor" because scientists are still not sure if it belongs to a larger family of receptors. GPR55 is widely expressed in the brain, especially in the cerebellum. It is involved in modulating blood pressure and bone density, among other physiological processes. GPR55 promotes osteoclast cell function, which facilitates bone reabsorption. Overactive GPR55 receptor signaling is associated with osteoporosis. GPR55, when activated, also promotes cancer cell proliferation, according to a 2010 study by researchers at the Chinese Academy of Sciences in Shanghai. This receptor is expressed in various types of cancer.

CBD is a GPR55 antagonist, as University of Aberdeen scientist Ruth Ross disclosed at the 2010 conference of the International Cannabinoid Research Society in Lund, Sweden. By blocking GPR55 signaling, CBD may act to decrease both bone reabsorption and cancer cell proliferation.

PPARs - nuclear receptor

CBD also exerts an anti-cancer effect by activating PPARs [peroxisome proliferator activated receptors] that are situated on the surface of the cell's nucleus. Activation of the receptor known as PPAR-gamma has an anti-proliferative effect as well as an ability to induce tumor regression in human lung cancer cell lines. PPAR-gamma activation degrades amyloid-beta plague, a key molecule linked to the development of Alzheimer's disease. This is one of the reasons why Cannabidiol, a PPAR-gamma agonist, may be a useful remedy for Alzheimer's patients.

PPAR receptors also regulate genes that are involved

in energy homeostasis, lipid uptake, insulin sensitivity, and other metabolic functions. Diabetics, accordingly, may benefit from a CBD-rich treatment regimen. CBD also exerts an anti-cancer effect by activating PPARs on the surface of the cell's nucleus.

CHAPTER SIX: CHOOSING THE BEST CDB OIL FOR YOUR NEEDS

Trying to find the best CBD oil can be very tough especially when you are just starting your research. Honestly, if you want to choose the best Cannabidiol or hemp oil and its products; you'll have to go through several channels and sometimes a variety of options in order to get correct information, strain and balance to help with the condition you are trying to alleviate.

Many people still confuse Cannabidiol (CBD) with Tetrahydrocannabinol (THC), the major psychoactive ingredient found in cannabis. Furthermore, CBD and hemp oil products are often confused to their origin, and medicinal properties. A large variety of CBD specific products have been legalized in a number of states because they do not give the high of THC, which many people run away from due to its psychological effects.

Many cannabis enthusiasts believe that the natural hemp extracts are the next generational as well as the next revolution in nutritional supplements. They believe that there are wider ranges of products that can help to deal with diseases fighting humanity in a natural and healthy way.

Things To Consider Before Choosing Your CBD Oil

CBD oil comes in various forms and products types. Choosing CBD oil can be overwhelming considering the fact that there are significant numbers of CBD products and brands in the market. Remember, CBD and THC work best together, enhancing each other's therapeutic benefits. For maximum therapeutic impact, choose products that include CBD and THC.

So with all these varieties in the market of today, which CBD oil is the best choice for you? The first step lies in knowing the way to compare products that are similar and differentiate those that seem identical. Therefore, to make an informed selection, these areas should be given adequate consideration:

CBD Oil Strength

Concentration of CBD in the product is the determining factor when choosing CBD oil products. Although the concentration can be misleading, it's the primary consideration when looking at edibles and digestible products.

Percent volume of CBD in the product - This property listing should be on the label and expressed as the percentage of the total volume of the product. It usually ranges from 0.1 to 0.26 in percentage. The concentration chosen would be dependent the amount of CBD you want to start with as well as the kind of product you would be sourcing it from.

Purity

It's important to consider what other property or thing your CBD oil has. The possibility of including preservation, additives, and solvents cannot just be ignored. Since CBD oil is derived from hemp - stalk, leaves, and flowers; and because farming methods are diverse, the possibility of CBD products contain pesticides, chemical fertilizers, and herbicides is 100% real. To avoid these unwanted chemicals and materials, seek products that are from natural, organic or well-tested synthetic industrial hemp sources.

Transparency

For you to determine the purity level of the CBD oil,

the original producers must be ready to certify lab analyses for every step of the product. This will help to show the concentration of CBD in the product as well as prove that the product is free from pesticides and other harsh chemicals. Look for products that are tested for consistency, and verified as free of mold, bacteria, pesticides, solvent residues, and other contaminants. Avoid products extracted with toxic solvents like BHO, propane, hexane or other hydrocarbons. Solvent residues are especially dangerous for immune-compromised patients. Look for products that entail a safer method of extraction like supercritical CO_2.

Price

The price of any product is influenced by its quality and purity. To produce high-quality CBD oil, a substantial amount of hemp is needed and an absolute refinement process is required as well. Select products with quality ingredients. No corn syrup, GMOs, trans-fats, and artificial additives. So, it's understandable that the purer and more concentrated the oil is; the higher the price will likely be.

This is the core reason why it is very important to look for a reputable company that will satisfy your CBD needs. Don't settle for poor quality oil just to save a few dollars, as most likely; you won't be getting any value.

CHAPTER SEVEN: WHAT IS PTSD?

PTSD is the abbreviation for "post-traumatic stress disorder," a mental health disorder that develops in some people that have witnessed or experienced a very traumatic event. Military veterans who have experienced combat, [12]people who have lived through massive natural disasters, and individuals who have experienced sexual trauma are the three main categories of individuals that develop PTSD, but these are not the only scenarios in which PTSD can develop.

For people who have experienced trauma in their lives, it can be normal to experience upsetting memories or even have trouble sleeping for a while after the event. Sometimes it can be hard to go to work, or attend school, or even spend time around those they care about greatly. However, most people when dealing with a traumatic loss or event eventually start to feel better after a few months. However, for people who develop PTSD, these night terrors and troubling memories become a new way of life.

Post-traumatic stress disorder can develop in anyone, and it is not a sign of weakness. While many different factors can increase the chance of it developing, many of those factors are not within a person's ability to control. Personal factors such as age, gender, and previous exposure to the type of trauma being experienced also affects whether the mental health

---

[12] https://www.ptsd.va.gov/professional/co-occurring/marijuana_use_ptsd_veterans.asp

issue develops or not, and in many cases can even determine how severe the PTSD will become once it rears its head.

The symptoms of post-traumatic stress disorder vary greatly, but they begin soon after the initial traumatic event that develops the condition. The key to identifying post-traumatic stress disorder is determining whether those symptoms eventually fade or not. If the symptoms become severe enough that it impedes on an individual's ability to operate in their own personal life, then it is a sign that PTSD has developed.

There are four main types of symptoms associated with PTSD that have a range of effects and a range of severity. Each person experiences their own symptoms in their own way, but all the symptoms and their levels of severity can be split into four main categories: reliving the event, avoiding situation that is a reminder of the event, feeling "keyed up", and developing more pessimistic feelings and beliefs.

Regarding reliving the event, this symptom comes into play in the form of memories or nightmares. Individuals feel as if they are going through the event again and can sometimes awaken at night sweating and crying out. However, memory-based flashbacks can also happen during the day if a sound or smell triggers a memory, and sometimes a person can lock-up in a public place, losing mobility because of the debilitating emotions cascading throughout their body.

This gives rise to the second category of symptoms and effects of PTSD, which is avoiding situations that serve as reminders of the triggering event. To avoid the first symptom of memories and nightmares, this is symptom is consciously or subconsciously employed. The individual withdraws from different situations that might have been enjoyable because he or she fears becoming "triggered," and the individual will even avoid new situations that he or she might enjoy before simply for fear of the flashback state. This category of symptoms also includes a certain degree of avoiding talk about the event, and it centers around the individual not wanting to be swept up in the emotional cascade of terrifying memory the event caused.

The third category is called feeling "keyed up," which is also known as hyperarousal. This means an individual suffering with PTSD might become jittery or seem to be on alert for some sort of danger. This might devolve into issues with sleeping or concentration and can begin to impede one's ability to focus and work. Those suffering with post-traumatic stress disorder might become easily angered or irritable, and they can develop unhealthy coping mechanisms, such as smoking or excessively drinking.

The fourth category of symptoms centers around a pronounced ability to hold pessimistic emotions and beliefs. The way sufferers of PTSD think about themselves might change because of the trauma they have experienced – they might start to feel shameful or guilty for things that do not merit shame or guilt. They might feel the entire world is a dangerous place and develop mistrust. Those struggling with PTSD who fall into this category of symptoms start finding it hard to feel happy.

Children can develop PTSD as well. The symptoms described above for adults can surface in children in much the same way, and as they grow and get older they will carry those symptoms with them. For example, children under the age of six might become very upset when their parents are not close by and they might experience erratic sleeping habits. Children between the ages of 7 and 11 then begin to act out the trauma through drawings or playtime or stories, and some might start exhibiting nightmares that further disrupt their sleep and require the nearby location of their parents. In this case, also, they might start to act out on their negative emotions. By the age of 12, children's symptoms of post-traumatic stress disorder surface more like adult symptoms: like anxiety, withdrawal, reckless behavior, and depression.

People who struggle with and suffer from post-traumatic stress disorder also experience other problematic issues – feelings of despair, drug problems, chronic pain that is not dedicated to one specific source, employment issues, and relationship problems are things that surface as a byproduct of the PTSD. In many cases, people who are dealing with these issues will also seek to treat them alongside the

PTSD because of their link to PTSD. Post-traumatic stress disorder has a negative stigma attached to it, so it is very common for it to go undiagnosed until someone begins to address some of the issues mentioned above. For example, a woman might not understand that she had PTSD from a traumatic sexual experience until her and her husband seek marriage counseling for their failing relationship.

The biggest question many people have regarding PTSD is whether it is curable. "Getting better" is defined differently for different people because not everyone experiences PTSD and its symptoms in the same way or with the same severity. There are many different standard treatment options for those who believe they are struggling with post-traumatic stress disorder, but none of them provide long-term relief. Some people, through time and exposure and therapy, can rid themselves of the most severe symptoms, but most people simply find that these things only help to dissipate the worst of the symptoms for a short period. In other words, standard treatment options help to abate, but not to cure or provide any long-term effects.

Many therapists who treat patients that deal with PTSD have the goal of simply helping them to obtain a specific quality of life. For those who struggle with PTSD, some part of their life usually becomes unnavigable. Maybe they have trouble holding down a job, or they struggle with continuing to cultivate their close social relationships. In this fashion, the post-traumatic stress disorder is beginning to impede their natural ability to function. Many therapists set their goals to ensure the patient can go back to operating their lives at a baseline level. They do not set a goal of eradicating the symptoms altogether.

The two main types of treatment prescribed for those who deal with PTSD are therapy and oral medication. For most people, the two are combined in order to find a balance. The therapy aspect simply involves meeting with someone and talking with them. The most effective talk therapy is called cognitive behavioral therapy, which is usually a mixture of cognitive therapy and exposure therapy. The cognitive therapy is where the individual is taught skills to help them understand how the trauma has since changed their baseline feelings and thoughts. The therapist helps them to wrap their mind around how their brain is now re-hardwired because of that event and how that event has changed how they feel and process inputted information.

The exposure therapy part is there the therapist encourages the patient to then begin talking out their trauma. They talk it out repeatedly, over and over again, until the memories that are triggered within their mind's eye are no longer upsetting to the patient. This helps the patient overcome their own thoughts and emotions regarding the trauma. It slowly enables them to assimilate themselves back into society because the memories being triggered no longer come with a need to withdraw from the world around them.

The issue with both of these treatment options, however, is the fact that a great deal of psychological and emotional pain comes with these coping and treatment techniques. For some, talking out a trauma that prompted such an intense psychological response can result in the emotional trauma becoming worse, especially if the person is attempting to suppress the worst of the event. For example, if someone is experiencing painful emotional flashbacks during the day because of abuse they suffered at the hands of a parent, they might be repressing a form of the abuse that took place. They might remember the times that their parent hit them, but maybe they are repressing the sexual trauma they endured as well. If things like this begin to surface in these types of therapy sessions, they can reset the process and, sometimes, spiral the patient out of control.

PTSD is a very serious mental health issue that millions of individuals around the world struggle with. It can be debilitating as it can cause a person to remove themselves from things in which he or she once found happiness, and the therapy and treatment processes outlined above can takes months, and even years, before they begin to have any positive effect on the individual. There must be another solution, right?

CHAPTER EIGHT: CAN CBD OIL HELP PTSD IN ANY WAY?

There is overwhelming evidence that suggests that CBD oil can help[13] those suffering with post-traumatic stress disorder. Government-funded studies around the world have spent hundreds of thousands of dollars studying the link between the cannabis plant and PTSD, and what was once seen as "cannabis abuse" is now being viewed as "cannabis coping."

The anxiety caused by this disorder can be attributed to a shift in the brain chemistry of an individual over time, usually when stress hormones and adrenaline-activated chemicals become hyper-responsive within the brain. This gives way to things like insomnia and issues with focusing on daily tasks, and fatigue can become something the PTSD sufferer battles on a regular basis. In many areas and states that still outlaw THC for medicinal and recreational purposes, people who struggle with post-traumatic stress disorder are forced to turn to things like antipsychotics and antidepressants.

These types of medications have been proven, time and time again, to have very little success in treating the more severe forms of PTSD. These drugs have substantial side effects, such as narcolepsy, that impede their users' abilities to hold onto jobs and simply drive vehicles. Many people are beginning to turn to CBD oil in order to help manage their post-

---

[13] http://herb.co/2017/04/23/cannabis-combats-ptsd/

traumatic symptoms. Researchers are starting to find that people struggling with PTSD have lower levels of anandamide in their system. This is an endogenous cannabinoid compound, and it is one that reacts with the receptors of the endocannabinoid system to help with a host of bodily issues. Anandamide is innate to all mammals, and it triggers the same receptors that are activated by THC when it is ingested.

Why is this important? When anandamide is not present in the levels in which it needs to be present, there arises an endocannabinoid deficiency. In other words, the body has stopped producing enough of its natural endocannabinoids to fill all the receptor sites necessary throughout the body, so the body can no longer trigger the natural therapeutic role of relaxation and calm. By providing the body the number of endocannabinoids it is missing, the therapeutic role of calm, relaxation, and control of emotions now becomes a reality.

In other words, CBD and THC do not remove a person's memories, but they provide the individual the endocannabinoids necessary in order to jumpstart the body's natural coping mechanisms. Instead of the hyperarousal symptoms and the highly-emotional states, the body can now regulate the individual's moods and mitigate his or her hyperarousal symptoms.

This endocannabinoid deficit physically contributes to the chronic anxiety, the impaired fear extinction, and the aversive memory consolidation that people struggling with PTSD experience. Another great aspect of CBD oil has no risk of overdosing or addiction. Not only that, but CBD oil has no major side effects. Some people might experience things such as dry mouth, but every other effect that comes with CBD oils and the strains of cannabis they are derived from are intentionally cultivated in order to produce those specific effects, and they will always be labeled and outlined.

Those who suffer with post-traumatic stress disorder are lacking in anandamide, and the side effect of that deficiency is an inability to cope with their mental health disorder. By giving the body what it needs [14]– that is, more anandamide – the body can begin to restore its natural balance and help ward off the severity of the symptoms being experienced.

Neurological Explanation Of PTSD

Looking deeper into the neurological causes of PTSD, we find that not only is posttraumatic stress disorder a result of anandamide deficiency, but it is also an inappropriate response of the brain to the environment. In a Michigan study[15] in 2014, researchers defined PTSD as the failure of the brain to inhibit inappropriate fear responses, relating it to other anxiety disorders like phobias and panic disorder.

[14] https://www.ncbi.nlm.nih.gov/pubmed/22736575
[15] https://www.ncbi.nlm.nih.gov/pubmed/23829364

A couple years earlier, in 2012, another study[16] was published in the ADAA Journal, defining PTSD as the persistence of aversive memories. That is, the researchers believed (and have been proven correct) that PTSD is a failure of the brain to forget or push past memories of danger even when no danger is presenting itself currently. This is also known as the failure of the extinction process. Specifically, the amygdala, which is associated with the fear response, becomes overactive, coupled with an irregulation of the prefontal cortical structures and abnormal functioning of the hippocampus and basal ganglia structures. This means that the centers for fear and memory are connecting dots where there are none to be connected, releasing a fear response throughout the body when no danger is present.

The endocannabinoid system is at play in this response, as it has been shown to be a modulator of adaptation to stress, regulating emotional memories in the amygdala and hippocampus. A study[17] in December 2013 studied survivors of the World Trade Center attacks and concluded that the endocannabinoid system regulates the memories we have that are attached to emotions.

A study[18] published in *Neuropharmacology* in January of 2013 found out more about fear extinction, which it defined as the decrease in a conditioned

---

[16] http://onlinelibrary.wiley.com/doi/10.1002/da.22031/abstract
[17] https://www.ncbi.nlm.nih.gov/pubmed/24035186
[18] https://www.ncbi.nlm.nih.gov/pubmed/22687521

response to a stimulus as the exposure to the stimulus is decreased, or "omitted." The study discovered that the endocannabinoid system and the glucocorticoid system both modulate the individual's emotional state and control emotional memory extinction. Thus, any fluctuation in the endocannabinoid or glucocorticoid systems could cause a fear response to pop up inappropriately. According to the research, fluctuations or interference with the endocannabinoid or glucocorticoid systems can cause inappropriate retention of memories, such that someone suffering from PTSD will experience resurfacing of frightening memories on a regular basis.

Another study[19] found that, specifically, the interruption of the CB1 receptor functioning decreases fear extinction in the individual. This means that situations that trigger memories continue to be terrifying to the individual who is suffering from PTSD. It contributes to the fact that sufferers of PTSD struggle with flashbacks and nightmares.

How do cannabinoids fit into this picture, then? Let's find out.

Studies On Cannabis And PTSD

A study[20] in July of 2013 in Brazil was published that spoke of the anxiolytic effects of cannabidiol. "Anxiolytic" means that the substance diminishes or effectively rids the individuals of anxiety. This

---

[19] https://www.ncbi.nlm.nih.gov/pubmed/15637635
[20] https://www.ncbi.nlm.nih.gov/pubmed/23433741

generalized effect helps many individuals who struggle with anxiety in various forms – phobias, panic disorder generalized anxiety disorder, and posttraumatic stress disorder. It effectively inhibits anxiety, depression, and psychosis, all of which can affect the individual with posttraumatic stress disorder.

The study we mentioned above from December of 2013 physically proved that there was a reduction in the circulation of endocannabinoids naturally within the body in individuals who developed PTSD following the World Trade Center attacks back in 2001. They found that endocannabinoid signaling had been identified as a means to modulate and adapt to stress, and found it to be a required trigger in the process of glucocorticoid regulation. The relationship and interactions between these endocannabinoids and the glucocorticoids have been shown to help regulate emotional memories, which means that when endocannabinoid signaling is impaired, it becomes harder and harder to regulate this facet of the human mind. This is when memories and emotions linked to those memories begin to rage out of control.

Participants in the study obtained a very structured diagnostic interview before being set up with their tests. Within these tests, they found physical evidence that PTSD is, in fact, associated with a reduction of circulation levels of certain endocannabinoids that are necessary to help emotionally regulate memories and their physical impacts.

Then, later in 2013, a study[21] was conducted that suggests that cannabinoids help to modulate the physiological and behavioral responses that the human body undergoes when experiencing stressful events or memories. When given cannabinoids, the brain circuitry that encompasses the hippocampus and the prefrontal cortex was boosted, which helped with contextual fear extinction. It was found that after one week of being regularly given cannabinoids, the patients began to experience their own strengthened abilities to emotionally regulate their physical responses to specific memories when triggered about their traumatic events, which suggests that the use of the cannabis plant and CBD oils can actually help those suffering with PTSD to not only cope with their mental disorder, but also help give them back bodily control over how they react to the memories when conjured. In other words, CBD oil doesn't just help to cope, it also helps to give back control.

Even more information exists pleading the case of cannabis versus posttraumatic stress disorder. A study[22] published in *Psychoneuroendocrinology* concluded, "Considerable evidence suggests that cannabinoids modulate the behavioral and physiological response to stressful events." In other words, cannabinoids like cannabidiol, or CBD, can affect and control our behavior and physical responses to events around us. This is because CBD is a CB1 and CB2 agonist, meaning it interacts with the

---

[21] https://www.ncbi.nlm.nih.gov/pubmed/23433741
[22] https://www.ncbi.nlm.nih.gov/pubmed/23433741

CB1 and CB2 receptors in a positive way, making single prolonged stress events (SPS events) less stressful and less memorable to the individual.

Neuropsychopharmacology published a study[23] in November of 2016 that cannabidiol, CBD, has pharmacotherapeutic properties. According to the scientists who performed the study, CBD interacts with mesolimbic dopamine production and serotonin production in the brain. By doing so, it blocks conditioned freezing behaviors. This could be revolutionary for sufferers of PTSD, as it means that memories triggering a panic response would no longer trigger panic if they were to use CBD oils or CBD in some other form.

In addition, another study[24] from December 2016, showed that cannabidiol reduced and even blocked the fear memory expression as well as promoted extinction, as shown by a later fear retention test. This is great news for those with PTSD, as it means that CBD can block the fear from arising with the memories of the traumatic experience such that fear is no longer attached to the memory. Without fear being attached to the memory, the memory will incite very little anxiety or panic. In turn, the individual with PTSD can venture out into life without the anxiety of a fear memory resurfacing and causing a negative response within himself or herself.

---

[23] https://www.ncbi.nlm.nih.gov/pubmed/27296152
[24] http://journal.frontiersin.org/article/10.3389/fphar.2016.00493/full

According to a 2007 study[25] published in *Molecular Neurobiology*, CB1 receptors, when activated correctly by endocannabinoids like anandamide or by external sources of cannabinoids, are associated with behavioral adaptation to aversive memories. In other words, ingesting, or consuming CBD can cause cannabidiol molecules to attach to the CB1 receptors in such a way as to modulate extinction learning in the brain. This causes the "aversive memories," that is, the memories of trauma, to have less effect and be less memorable to the to individual.

Differences Between SSRI's And Cannbinoids

Here is the main difference[26] between CBD oil treating post-traumatic stress disorder and everything else on the market that is prescribed: CBD oil helps with the modulation of the fear conditioning process. When someone is diagnosed with PTSD, selective serotonin reuptake inhibitors (SSRIs) are usually the first course of action in the treatment of this ailment. However, many patients that begin down this road end up battling their same symptoms again while they explore other pharmacological agents to mitigate the other symptoms that the SSRIs are not targeting. This leads to a massive risk of mixing improper medications as well as accidental overdose, and the lack of being able to mitigate and modulate specific

---

[25] https://www.ncbi.nlm.nih.gov/pubmed/17952654
[26]https://www.ncbi.nlm.nih.gov/pubmed/?term=traumatic+and+K erbage+H

symptoms associated with PTSD is what ultimately brings about suicidal thoughts.

While many different drugs and pharmacological agents have been tested in many studies and have shown a general positive short-term effect for a patient suffering with PTSD, the medication gives them no hope of long term mediation and modulation when it comes to their fear conditioning. In other words, it gives them no control back over their emotional states that are triggered because of their memories.

For this reason, many people trying to cope with post-traumatic stress disorder turn to CBD oil and the cannabis plant in order to help treat them medically – because the long-term effects of these cannabinoids are actually giving them back control over their bodies, emotions, and memories. This idea of extinction of fear exists when PTSD is treated with CBD oil, but it does not exist at all when treated with regular medications and standard therapy practices. This fear extinction is what ultimately improves the core symptoms of those suffering with PTSD, and fear extinction is the road that ultimately helps give hope to those suffering that, one day, they might be rid of their PTSD altogether. Pharmacological drugs do not give them that hope. CBD oil and other cannabinoids can be and are being proven via the same science that gave us SSRIs in the first place.

Those who have started using CBD oil as a way to medically treating their PTSD have boasted[27] of better moods, better quality of sleep, improved memory and thinking, and better overall performance in their work environment as well as their social life. This is because of the reaction that takes place when the CBD oil gives the body the missing endocannabinoids it needs to function properly. Think of it this way – if your body is deficient in vitamin D, the biggest symptom that is experienced is depression. Thus, those who battle depression during the winter months are told to take a high-dose vitamin D supplement every single day. Their body is lacking in an essential vitamin, and that deficiency is contributing to their depression.

One particular case[28] of a nineteen-year old male with severe PTSD reports that the individual was experiencing major panic attacks and intense flashbacks and even engaged in self-mutilation. This individual's PTSD was raging out of control until he was introduced to cannabis as a solution. He began to calm as soon as the cannabis was consumed, and results indicated that the stress-sensitive nuclei in the hypothalamus had been affected by the traumatic event in the individual's life and that the cannabinoids present in the cannabis were helpful in repairing those nuclei and returning them to their healthy state. Thus, it was shown that cannabinoids are a means of reparation and healing to the brain that has been traumatized and has developed posttraumatic stress disorder.

---

[27] https://elixinol.com/blog/cbd-oil-ptsd-post-traumatic-stress-disorder/

[28] https://www.ncbi.nlm.nih.gov/pubmed/22736575

CBD oil is being proven to be a tremendous help to those dealing with post-traumatic stress disorder. It is starting to give a massive community of people hope that, one day, they will regain control over their own lives and be the controller of their own emotional states once again. The idea of fear extinction is providing hope that they will be able to eradicate their symptoms altogether and not allow their traumatic experience to haunt their lives any longer.

CBD oil can most certainly help those dealing and coping with PTSD, and it is simply a matter of figuring out how you wish to deliver the cannabinoids into your system. There are multiple products that have a variety of delivery methods attached to them, and it is all based on what makes you comfortable and the dosage amount you want to take in daily.

CHAPTER NINE: PRODUCTS AND BENEFITS IN TREATING PTSD WITH CBD OIL

As we have already discussed, studies over the past few years have given rise to the notion that CBD oil can help with PTSD. Many states in the U.S. have legalized the medical use of marijuana, but there is still a very negative stigma that comes with the cannabis plant in its entirety. However, it has also been proven to help tremendously with post-traumatic stress disorder and the issues that come along with it, while precious, valuable members of our community are reaping the negative consequences of not being able to utilize cannabinoids as a viable medical option, specifically, the largest part of the population that suffers with PTSD – soldiers and war veterans.

This mental condition is usually accompanied by hypervigilance, or the enhanced feeling that a threat is lurking around every corner. This can lead to all sorts of issues, such as adrenal glandular burnout and insomnia, and these effects can themselves have life-threatening consequences if not taken care of properly. PTSD can wreak havoc on an individual's ability to function in their normal, everyday life, and it can bring on different types of anxiety-based attacks that can put an individual in the hospital if he or she is not careful.

Post-traumatic stress disorder has the highest drug abuse percentage of any mental illness out there today. The flashbacks and the emotions that come with them are too heavy and frightening to bear, so the only means by which an individual feels relief is possible is by abusing things like alcohol and prescription drugs. It is a desolate mindset that can completely decimate someone's life, but cannabinoids and CBD oil can offer another option for those suffering with this mental illness. CBD oil and cannabinoids can give them the ability to get on with their lives without the feeling of those troubling symptoms, and can even free them of the "triggering" that comes with the memories of the event completely.

Increasingly, those who serve in militaries across the globe are beginning to experience the devastating mental and physical effects of PTSD. Their exposure to modern warfare and the atrocities that come with it is tainting the minds of individuals from those in their teenage years up to adults in their fifties. These violent combat situations are completely changing the chemical makeup of their brains, and it is resulting in an inability to assimilate normally into society as well as to their job in the military. Those who suffer with PTSD are usually medically discharged from the entity that exposed them to this risk in the first place, at which point one of the main positive treatment options, cannabis, is forbidden by law or by social stigma, or by both.

CBD and other cannabinoids have no risk of dependency, addiction, or overdose, and they can be mixed or taken alongside anything. It is the perfect medication for someone battling something like PTSD, and yet this regimen is forbidden to PTSD sufferers because it could "get them high." The tormenting mental and physical walls that come with post-traumatic stress disorder are contributing to the rising rate of suicide among soldiers and war veterans. So many deaths could be avoided and the suffering of PTSD could be eliminated if CBD oil as well as other cannabinoids could be freely used. The rate of soldier and veteran suicide, as of 2016, is 23 deaths per day, and that number will continue to steadily rise unless we find a way to eliminate the debilitating issues that come with this type of mental condition.

CBD oil helps to reduce the concept of fear memory while removing the high levels of THC in the cannabis plant. This helps reduce anxiety levels while still allowing a veteran or soldier to obtain their VA and medical benefits, since if psychoactive cannabinoids are found within their system they could lose their benefits. For this reason, CBD oil is a tremendous help to those suffering with PTSD, and especially with soldiers – CBD oil gives them relief they cannot find with traditional medications while still enabling them to keep the benefits they so deserve.

Around 8 million Americans, not including those that serve in the military, suffer with PTSD on a regular basis because of some traumatic event they endured during their lifetime. There are no effective, approved, or even specialized treatments for those suffering from this mental illness, yet something like CBD oil that has scientific advancements and research behind it is still being withheld from them. Instead, they are being treating with psychiatric drugs that induce things like paranoia and suicidal thoughts – the exact symptoms people who suffer with PTSD are attempting to get away from.

The neurobiology of post-traumatic stress disorder sufferers follows this pattern – a neurotransmitter imbalance becomes apparent in the brain that affects serotonergic and noradrenergic reactions and mechanisms within the brain. These neuroanatomical disruptions that take place ricochet into the immune system and the autonomic nervous system, and this leads to the imbalance that causes PTSD. But, this involvement of multiple factors presents an issue when single-molecule psychiatric drugs come into play, because no one oral medication is formulated to treat all of those imbalances in all those systems. However, the endocannabinoid system within the human body spans all of those systems as well as those areas of the brain affected by PTSD, which means ingesting cannabinoids can be a theorized medication that targets all of these faculties at once.

The endocannabinoid system also plays an important role in the brain's prefrontal cortex functionality. This part of the brain is involved in information processes, sexual arousal, and controls conditioned fear responses. The endocannabinoid receptors found scattered throughout the prefrontal cortex, when activated, can be involved in disrupting that conditioned fear response, resulting in alleviation from the major symptoms of PTSD. This could then decrease the suicide rate among those suffering with post-traumatic stress disorder, and it would give soldiers and war veterans the capability of leading a fully-functioning life and the ability to stay in the military, should they choose to do so.

Because cannabinoids also have no risk of addiction or overdose, those suffering from PTSD have the ability to take this drug as many times as they need whenever they feel they need it without wondering if they are becoming dependent or going to take too much. Another great thing about this is that CBD oil and its related products can be distributed in a variety of ways, from smoking to diffusing essential oils within the home. CBD tinctures can be put into foods and teas to be ingested, skin patches can be adhered to the skin, CBD shampoo can be used during regular grooming activities to help with a boost throughout the day, and CBD lotions and balms can be applied topically and rubbed into the skin.

CBD and other cannabinoids can be inhaled for a rapid reaction, rubbed into the skin for a slow-release reaction, and infused throughout the home via essential oil diffusers for an all-day dosage. However, choosing the product that is right for you requires looking into a few different things, such as the concentration of the dosage, the number of doses in a day, and the percentage of concentration of CBD in the oil that is purchased.

PTSD has no specialized treatment options for those who suffer with it. There is no tailored medication for their ailment, there is no step-by-step program to cope. There are no standardized psychologists who specialize in this type of mental disorder. For something that plagues millions upon millions of individuals a day, one would assume steps have been taken to regulate and standardize some sort of treatment plan. However, few steps have been taken at all, and the steps have been uncoordinated at that. This leaves tens of millions of individuals across the globe suffering silently while the an ideal form of medication that could provide relief remains out of their reach.

Scientific studies are emerging that show us exactly how useful CBD oil and its related cannabinoids and products can be, and physicians and psychologists alike are beginning to do their own research and attend seminars in an attempt to understand how the substances could help their patients. Some doctors are even beginning to prescribe this type of treatment within their own offices. The one factor that must be taken into consideration is the THC content of the oil or related product being purchased. Some extraction methods to obtain the CBD oil completely remove THC from the mixture, while other extraction methods leave decent amounts of THC within the resin byproduct. If you live in an area where THC is outlawed for any reason, keep watch for the CBD to THC ratio in the products you research.

Understanding what a strain is will aid in your understanding of those percentages and ratios. When someone talks about a particular cannabis strain, what do they mean? IN a legal cannbais distribution center, there are clear glass containers arranged in different orders with names on them like "White Kush" and "Super Sour Diesel," which denote the strain of cannabis inside. They each have specific side effects and reactions within the body, and some are better to take at night while some are formulated for daytime use.

So, how do you know which one is right for you? Do not worry, we will outline all of this below as well as give you the top five cannabis strains that are the most beneficial in treating PTSD.

CHAPTER TEN: THE BEST STRAINS FOR TREATING PTSD

Before we talk about the most popular strains that will benefit PTSD sufferers, it is important to know that a strain means the specific parent plant from which the bud was bred. Cannabis strains are specifically cultivated to produce certain effects, and they can even been grown and bred with a specific ratio of certain cannabinoids like THC and CBD.

However, there are strains that are more beneficial than others when it comes to treating specific medical conditions. If you are someone who struggles with sleeping, you do not want something that is going to give you energy. That rush of energy will only serve worsen your symptoms. In this way, the strain of cannabis you choose can be tailored to help you with your sleeplessness and reduce your insomnia episodes through a relaxing or sedative side effect.

Strains that are beneficial for those dealing with PTSD are strains whose effects are soothing and target that conditioned fear response. However, there exist some strains that help with the other symptoms of PTSD, such as sleeplessness due to memories and nightmares, the anxiety induced by flashbacks, and even the flashbacks themselves.

When researching strains, it's important to understand the type of seed the cannabis plant came from. In this category, there are four different kinds of seeds cultivators use, and each produce variations of the cannabis plant. They come with different physical appearances, undertones of taste, and side effects induced when ingested, and knowing the type of seed your cannabis comes from can help you nail down with consistency your product once you find the one that works best for you.

The most ideal seed type is called the clone-only seed variety. Growers who use this type of seed take genetic clones of the original plant they want to grow and cultivate it closely and with intense fervor over a period of several plant generations. This helps to keep a stable cannabis strain, so the side effects, looks, taste, and origin will always be the same. This is the most desirable seed type from which to obtain your CBD oil because it can provide the most consistent product variables when you are reordering from a specific manufacturer.

Another beloved seed type is called a stable seed variety. This means the grower is being incredible selective in choosing the seeds he or she wants from each plant generation. Then, the grower takes those ideal seeds and continues breeding them over a period of multiple generations in order to get the exact outcome they are looking for. This is usually how hybrid strains are grown, and indeed is the ideal way for hybrids to be bred. Think of it as "forced Darwinism" for cannabis strain growing and cultivating. When the final product that the grower desires is reached, those seeds are then taken and used to grow the rest of the grower's crop.

Another type of seed is called the unstable variety seed, but this is not an ideal seed from which to harvest your regular product. The buds from strains that grow from these plants are grown without testing different generations of seed. This can benefit the grower with faster production because of the lack of a need to grow multiple generations of plants, but the drawback is that these seeds will grow cannabis that comes with uneven characteristics. Thus, one CBD oil product you purchase from someone using these unstable seeds might work wonders, but the next product you receive might be sub-par and doesn't deliver the same potency or side effects you were seeking. In general, stay away from products and buds using this type of seed.

The last seed type is not really a type but rather is the original cannabis plant. It is called a landrace variety (or "wild varieties"), and these are the cannabis plants growing in the wild across the world. These wild-growing crops are the original parent plants of all the cannabis in circulation around the world in areas where it does not naturally grow. If you can find out the original, wild-grown landrace of your strain, then you have the best piece of information to help you single out strains of cannabis that will ultimately help in medicating your PTSD.

Below are the five most popular strains of cannabis for those suffering with PTSD. This list is only to get you started. Keep a running document of all the strains you experiment along with their side effects and effectiveness. Then, see how much information you can discover about them – list the company who sold you the product, the grower of their product, the origin of their product, and reach out to the grower yourself and see if you can figure out what types of seeds they use when cultivating their cannabis plants. All this information will help you locate the exact product that works the best for you, and it will guide you when reordering the CBD oil you are looking for.

Ringo's Gift

The first strain is called "Ringo's Gift," and this hybrid has a 24:1 ratio when it comes to its CBD to THC content. This strain possesses an incredible relaxation effects as well as the side effect of a mellow cerebral experience. This can be wonderful in calming those who frequently experience panic and anxiety attacks due to their flashbacks. The relaxation and calming effects are not sedative, however, so this is a wonderful strain that can be used throughout the day to help whenever someone feels the onset of a panic episode beginning to occur. However, many sufferers of PTSD also keep it by their bedside if they are awoken by their nightmares, as this strain helps them to settle down and get back to sleep.

ACDC

The next strain is called "ACDC" and is another high CBD strain with a CBD to THC ratio of 20:1. The high CBD content induces a calm mind and promotes relaxation, which can be beneficial for PTSD-sufferers who have problems sleeping or become anxious throughout their day. It also helps to treat nausea, which is a common side effect of those struggling with PTSD due to the emotional trauma that can be brought on during triggered flashbacks. The calm mind this strain promotes will slowly help to ease any PTSD sufferer back into society by helping to lower the chances of a triggered outbreak happening in public.

Cannatonic

Another popular strain is called "Cannatonic," which promotes not only relaxation but also an uplifting mindset. Cannatonic can be very beneficial to PTSD sufferers who battle overall aches and pains due to a plethora of reasons, from lack of sleep to panic attacks. This strain can help ease those tense muscles as well as get rid of those muscle spasms, and the relaxation it induces can help with the panic and anxiety that arises during triggered episodes that bring on the muscle tension in the first place. This is a wonderful all-day strain to have around because of the sheer number of symptoms it can treat, and it can do wonders to help someone with PTSD feel their best.

Harle-Tsu

Another great strain for PTSD sufferers is a strain called "Harle-Tsu." This hybrid strain has a CBD content that is at least 20 times higher than its THC content. While this strain has anti-inflammatory effects, it also has analgesic effects as well as gives the patient a healthy dose of relaxation. It promotes a calm mindset to help with the fear conditioning a PTSD sufferer has suffered, and if any pain is being experienced because of anxiety or panic attack, day or night, then this strain is the perfect blend for you. The relaxation is not sedative, so it is a good strain to help upkeep one's mental disposition throughout the day, and the calm mindset it induces helps a PTSD sufferer to continue going out into public. With this strain, an individual suffering from PTSD becomes no longer a slave to their anxiety and the physical and muscular pain that accompanies anxious symptoms.

Charlotte's Web

This last strain is called "Charlotte's Web." It should sound familiar because we discussed it earlier in this book. The pain relief and relaxation that comes with this strain has branded it one of the most popular strains of high-CBD cannabis on the market. It has been used to treat everything from PTSD to epilepsy, and its potency is incredible. It induces a relaxed sensation as well as an uplifted mindset, which can do wonders for those suffering with PTSD. However, other side effects, such as an increased focus in daily activities, has pushed this strain over into the PTSD world. Its ability to focus the mind makes this the perfect strain for those suffering with PTSD to maintain their jobs, keep up with their familial lives, and even help those who have become housebound because of their condition.

SUMMARY

CBD oil is most certainly effective when it comes to treating post-traumatic stress disorder. Those who are dealing with many mental illnesses turn to cannabis and CBD oil in order to alleviate their symptoms. From those suffering from anxiety to those who struggle with depression and suicidal thoughts, many people have taken to the internet and told the public (as well as their doctors) how cannabis and CBD oil have helped them on their journey to recuperate and take back their lives. Sufferers of PTSD are not strangers to anxiety and depression, and suicide rates among military service members struggling with PTSD is some of the highest in the United States. This is a problem that has no traditional fix, but CBD oil *can* help solve the problem.

Because of the way cannabinoids interact with the brain and the receptors of the endocannabinoid system, ingesting outside sources of cannabinoids can help to fill holes the body has created in its own chemical processes as a result of the traumatic event the individual experienced. Just as we take in outside sources of dopamine or serotonin in an attempt to traditionally regulate the symptoms of depression, when there is a gap in chemical messages in the endocannabinoid system the ingestion of outside sources of cannabinoids helps to fill those gaps.

The issue is that the entire cannabis plant has been stigmatized because of a few psychoactive agents, which can be bypassed with certain extraction methods. PTSD has no traditional treatment paths once it is diagnosed. Some might seek therapy, some might attempt to take psychiatric drugs whose side effects are the very things the patient is trying to have alleviated, and some begin to abuse addictive substances in order to cope with the anxiety that comes with their hypervigilance. Meanwhile, CBD oil has been proven to promote anti-anxiety effects as well as rebalance the chemical messages being sent through the endocannabinoid system. It can be effective without giving the "high" of THC that is so stigmatized.

Not only can CBD help, but there are a variety of options for delivery to the body. From CBD sprays to CBD essential oils diffused throughout the home, this is a medication that can be specifically tailored to a delivery method that is not just the oral ingestion of a pill. Not only that, but it has no risk of overdose and addiction and it poses no risk of fatally mixing with another substance. It can be used in the replacement of oral drugs or it can be used alongside without issue, and it can be taken in as many doses over the course of the day as the patient needs in order to help them cope with their post-traumatic stress.

CBD oil is not psychoactive, so it will not induce the "high" people associate with the cannabis plant. This means that using CBD oil will not put soldiers and war veterans at risk of losing their benefits or their jobs when all they want to do is effectively relieve symptoms of their mental illness. Avoiding these risks while obtaining the positive effects of adding a cannabinoid regimen to one's life could turn a person's life around for the better – medical marijuana and CBD oil products can literally save someone's life.

The scientific research and the personal stories that pepper the internet and the scientific community are growing by the dozens every single day. For a mental illness like PTSD that has no standard level of treatment and no action-plan when it comes to coping, CBD oil could be the first-ever standard by which something like this is treated. It can turn entire families around for the better and it can keep children from losing their parents to unnecessary actions of suicide. In theory, the use of CBD oil to help treat and cope with PTSD could even reduce the overall drug and alcohol addictions across the world!

If you or someone you know is suffering with PTSD, CBD oil is a viable and legal way for them (or you) to begin treating the debilitating symptoms of this mental illness. In a medical society that has let you down, CBD oil can help bring you back up in ways you cannot currently imagine. It has no risk of overdose, it has no addictive properties, and it can be taken alongside any other medication on the market. It is worth a shot, and it is a shot you will not regret.

And here is your FREE BONUS MATERIAL![29]

---

[29] https://steemit.com/@cbdjane

Made in the USA
Columbia, SC
08 July 2018